# Grid Modernization: One Grid or Many?

[*pilsa*] - transcriptive meditation

**AI Lab for Book-Lovers**

*synapse traces*

xynapse traces is an imprint of Nimble Books LLC.
Ann Arbor, Michigan, USA
http://NimbleBooks.com
Inquiries: xynapse@nimblebooks.com

Copyright ©2025 by Nimble Books LLC. All rights reserved.

ISBN 978-1-6088-8415-5

Version: v1.0-20250830

*synapse traces*

# Contents

| | |
|---|---|
| Publisher's Note | v |
| Foreword | vii |
| Glossary | ix |
| Quotations for Transcription | 1 |
| Mnemonics | 183 |
| Selection and Verification | 193 |
|     Source Selection | 193 |
|     Commitment to Verbatim Accuracy | 193 |
|     Verification Process | 193 |
|     Implications | 193 |
|     Verification Log | 194 |
| Bibliography | 205 |

*Grid Modernization: One Grid or Many?*

# Publisher's Note

At xynapse traces, we observe the critical systems that underpin human civilization. Few are as foundational as the energy grid, the circulatory system of modern life. The question of its future—a single, monolithic network or a resilient mesh of distributed nodes—is not merely technical; it is a profound inquiry into our models for community, autonomy, and ecological balance. The data streams on this topic are vast and often contradictory, flowing from dense engineering studies to the speculative landscapes of fiction. To truly comprehend the stakes, a different mode of processing is required.

This is the purpose of 필사 (p̂ilsa), or transcriptive meditation. By slowly and deliberately transcribing the words of engineers, planners, and visionaries, you engage in a unique form of cognitive mapping. The act of writing by hand bypasses passive consumption, allowing complex ideas to root themselves in your understanding. You begin to perceive the intricate logic, the hidden assumptions, and the ethical weight within each perspective. This collection is not simply a book to be read, but a meditative tool designed to help you internalize the architecture of our potential futures. Through this practice, you are not just learning about the grid; you are tracing the very blueprints for human thriving in an uncertain world.

*synapse traces*

# Foreword

The act of transcription, known in Korea as 필사 ( p̂ilsa), is a practice that transcends mere mechanical reproduction. It is a profound and deliberate engagement with the written word, a form of intellectual and spiritual discipline with deep roots in Korean history. To view p̂ilsa as simple copying is to miss its essence; it is, rather, a slow, contemplative dialogue between the reader, the author, and the very fabric of the text itself. By tracing the author's thoughts with one's own hand, the transcriber internalizes not only the content but also the rhythm, structure, and spirit of the work.

Its origins are embedded in the peninsula's dual scholarly traditions. For the Confucian scholar-officials of the Joseon Dynasty, the seonbi
(선비), p̂ilsa was an essential pedagogical tool for mastering the classics, ensuring that the ethical and philosophical tenets of the texts were absorbed into one's character. In parallel, within Buddhist monastic life, the practice of sutra copying, or 사경 (sagyeong), served as a powerful meditative act—a devotional method for cultivating mindfulness, purifying the mind, and accumulating merit. In both contexts, the physical act of writing was inseparable from the mental and spiritual process of learning and contemplation.

With the advent of mass printing and the accelerated pace of modernization, the slow craft of p̂ilsa inevitably declined. Yet, in a compelling paradox, this ancient practice is experiencing a remarkable resurgence in our hyper-digital age. As our lives become increasingly saturated with fleeting digital content and ephemeral communication, many are turning to p̂ilsa as an analog sanctuary. It offers a tangible antidote to digital fatigue, a quiet rebellion against the culture of speed and distraction.

This revival illuminates a fundamental human need for deeper connection. For the modern reader, p̂ilsa transforms the passive consumption of text into an active, embodied experience. It forces a deceleration, demanding a level of attention that fosters concentration and enhances comprehension. In an era of information overload, the simple, focused act of putting pen to paper offers a path back to the quiet center of the reading experience, reminding us that to truly understand a text, we must sometimes inhabit it, one character at a time.

# Glossary

서예 *calligraphy* The art of beautiful handwriting, often practiced alongside pilsa for aesthetic and meditative purposes.

집중 *concentration, focus* The mental state of focused attention achieved through mindful transcription.

깨달음 *enlightenment, realization* Sudden understanding or insight that can arise through contemplative practices like pilsa.

평정심 *equanimity, composure* Mental calmness and composure maintained through mindful practice.

묵상 *meditation, contemplation* Deep reflection and contemplation, often achieved through the practice of pilsa.

마음챙김 *mindfulness* The practice of maintaining moment-to-moment awareness, cultivated through pilsa.

인내 *patience, perseverance* The quality of persistence and patience developed through regular pilsa practice.

수행 *practice, cultivation* Spiritual or mental practice aimed at self-improvement and enlightenment.

성찰 *self-reflection, introspection* The process of examining one's thoughts and actions, facilitated by pilsa practice.

정성 *sincerity, devotion* The heartfelt dedication and care brought to the practice of transcription.

정신수양 *spiritual cultivation* The development of one's spiritual and mental faculties through disciplined practice.

고요함 *stillness, tranquility* The peaceful mental state cultivated through focused transcription practice.

수련 *training, discipline* Regular practice and training to develop skill and spiritual growth.

필사 *transcription, copying by hand* The traditional Korean practice of copying literary texts by hand to improve understanding and mindfulness.

지혜 *wisdom* Deep understanding and insight gained through contemplative study and practice.

*synapse traces*

# Quotations for Transcription

The following quotations are offered for transcription, a practice of mindful engagement with complex ideas. As you manually copy these words—whether from a dense engineering study or a piece of speculative fiction—consider the parallels between your task and the subject at hand. The act of transcription requires a slow, deliberate focus, forcing you to trace the architecture of an argument, much like an engineer traces the circuits of a grid. You are not merely consuming information; you are actively building it, connecting word to word, concept to concept.

In this quiet, distributed act of personal study, you engage with the grand, centralized plans and decentralized visions that shape our energy future. Let your pen or keyboard be a tool for connection, a way to internalize the intricate details of grid modernization. Through this focused practice, you can transform abstract infrastructure plans into tangible, personal understanding, one letter at a time.

The source or inspiration for the quotation is listed below it. Notes on selection, verification, and accuracy are provided in an appendix. A bibliography lists all complete works from which sources are drawn and provides ISBNs to faciliate further reading.

[1]
> *High-voltage transmission lines are the arteries of our nation's energy system, and building more of them is essential to adding more clean energy to the grid, improving resilience, and lowering costs for American families.*
>
> U.S. Department of Energy, *Building a Better Grid Initiative* (2022)

*synapse traces*

Consider the meaning of the words as you write.

[2]

> *Substation modernization is driven by the need to replace aging infrastructure, improve reliability and resiliency, lower operating costs, and meet new digital requirements for data and security. The digital substation is a key part of this evolution.*
>
> ABB Group, *The Digital Substation: A key element of the smart grid* (2017)

*synapse traces*

Notice the rhythm and flow of the sentence.

[3]

> *Historically, the inertia from the rotating mass of synchronous generators was the primary source of frequency stability. As inverter-based resources (IBR) displace synchronous generators, the total inertia of the system declines, making the grid more vulnerable to frequency disturbances.*
>
> National Renewable Energy Laboratory (NREL), *Inertia and the Power Grid: A Guide Without the Spin* (2020)

*synapse traces*

Reflect on one new idea this passage sparked.

[4]

*The average age of the installed base of LPTs is approximately 40 years, which approaches the typical design life of 30 to 40 years.*

U.S. Department of Energy, *Large Power Transformer Study* (2023)

*synapse traces*

Breathe deeply before you begin the next line.

[5]

*The increasing complexity of the grid, with its intermittent renewables and distributed energy resources, makes this more challenging than ever.*

Pacific Northwest National Laboratory (PNNL), *Human-in-the-Loop Simulation for Power Grid Control Centers* (2021)

*synapse traces*

Focus on the shape of each letter.

[6]

*Much of our nation's energy system predates the 21st century. Most of our electric transmission and distribution lines were constructed in the 1950s and 1960s with a 50-year life expectancy, and are in need of replacement.*

American Society of Civil Engineers (ASCE), *2021 Report Card for America's Infrastructure: Energy* (2021)

*synapse traces*

Consider the meaning of the words as you write.

[7]

*Under traditional rate-of-return (ROR) regulation, regulators set rates so that the utility has an opportunity to recover its prudently incurred costs of providing service, including a return on its capital investment.*

National Association of Regulatory Utility Commissioners (NARUC),
*An Introduction to Regulation* (1970)

*synapse traces*

Notice the rhythm and flow of the sentence.

[8]

*Organized wholesale electricity markets are intended to foster competition among electricity generators, which helps to reduce the cost of electricity for consumers. These markets use a bidding system to determine which generators will produce electricity and at what price.*

Federal Energy Regulatory Commission (FERC), *Understanding Wholesale Electricity Markets* (2022)

*synapse traces*

Reflect on one new idea this passage sparked.

[9]

> *Stranded assets are assets that have suffered from unanticipated or premature write-downs, devaluations or conversion to liabilities.*
>
> Carbon Tracker Initiative, *Stranded assets: a climate risk challenge* (2017)

*synapse traces*

Breathe deeply before you begin the next line.

[10]

*Generally, FERC has jurisdiction over wholesale sales of electricity and transmission of electricity in interstate commerce, while states have jurisdiction over retail sales and local distribution.*

Congressional Research Service, *Jurisdiction and Regulation of the U.S. Electric Grid* (2022)

*synapse traces*

Focus on the shape of each letter.

[11]

*A key debate in transmission planning is how to allocate the costs of new high-voltage lines. Should the costs be borne by the generators who benefit, the customers in the region receiving the power, or spread more broadly?*

American Clean Power Association, *Transmission Cost Allocation* (2021)

*synapse traces*

Consider the meaning of the words as you write.

[12]

> *The traditional utility business model is a natural monopoly, where a single company owns and operates the generation, transmission and distribution infrastructure in a specific service territory. ... This traditional structure is being challenged by a number of factors, including the rise of distributed energy resources (DERs) such as rooftop solar and battery storage.*
>
> Smart Electric Power Alliance (SEPA), *The Changing Utility Business Model* (2018)

*synapse traces*

Notice the rhythm and flow of the sentence.

[13]

> *The U.S. power grid is vulnerable to physical attacks on critical infrastructure, such as substations and transmission lines. According to some studies, a coordinated attack on a few key substations could potentially lead to a widespread, long-duration blackout.*

Congressional Research Service, *Physical Security of the U.S. Power Grid* (2022)

*synapse traces*

Reflect on one new idea this passage sparked.

[14]

*Cybersecurity threats to the electric grid are growing and becoming more sophisticated. ... By exploiting vulnerabilities in grid systems, attackers could disrupt grid operations, cause equipment damage, or trigger power outages.*

U.S. Government Accountability Office (GAO), *Electricity Grid Cybersecurity: DOE and FERC Need to Take Further Action to Identify and Address Risks* (2021)

*synapse traces*

Breathe deeply before you begin the next line.

[15]

*Blackstart capability is the procedure to restore an electric grid to operation without relying on the external electric power transmission network to recover from a total or partial shutdown.*

North American Electric Reliability Corporation (NERC), *Blackstart and Next Start Resource Issues White Paper* (2020)

*synapse traces*

Focus on the shape of each letter.

[16]

*Extreme weather events, which are increasing in frequency and intensity due to climate change, are a leading cause of power outages in the United States.*

The White House Council on Environmental Quality, *Opportunities to Advance Equity in Utility Regulation: A Report to Congress* (2021)

*synapse traces*

Consider the meaning of the words as you write.

[17]

*The removal of a single node in these networks can lead to a cascade of failures that serves to break them into many disconnected islands.*

A. E. Motter and Y.-C. Lai, *Cascade-based attacks on complex networks* (2002)

*synapse traces*

Notice the rhythm and flow of the sentence.

[18]

*A severe GMD event could cause widespread and long-duration blackouts by inducing powerful, uncontrolled currents in the electric transmission grid, which can damage or destroy large power transformers.*

Federal Energy Regulatory Commission (FERC), *Geomagnetic Disturbances* (2013)

*synapse traces*

Reflect on one new idea this passage sparked.

[19]

> *At the end of 2022, over 2,000 GW of generation and storage were actively seeking grid interconnection... The total capacity in the queues is now ~1.5x the total installed capacity of the U.S. power fleet (~1,250 GW).*
>
> Lawrence Berkeley National Laboratory, *Queued Up: Characteristics of Power Plants Seeking Transmission Interconnection, 2023 Edition* (2023)

*synapse traces*

Breathe deeply before you begin the next line.

[20]

*Curtailment is the deliberate reduction in output of a generator from what it could otherwise produce, typically to balance energy supply and demand or to manage transmission congestion.*

Stoel Rives LLP, *Renewable Energy Curtailment: A-to-Z* (2020)

*synapse traces*

Focus on the shape of each letter.

[21]

*Effective coordination between adjacent balancing authorities is essential for maintaining grid stability. This includes sharing real-time data on generation, load, and interchange to ensure the system remains in balance across wide geographic areas.*

North American Electric Reliability Corporation (NERC), *NERC Reliability Standards (e.g., BAL-002)* (2019)

*synapse traces*

Consider the meaning of the words as you write.

[22]

*Ancillary services are functions that help grid operators maintain a reliable electricity system.*

U.S. Energy Information Administration (EIA), *Today in Energy* (2021)

*synapse traces*

Notice the rhythm and flow of the sentence.

[23]

*In this way, storage can act as a 'shock absorber' for the grid, smoothing out the fluctuations in energy supply and demand.*

Union of Concerned Scientists, *How Energy Storage Works* (2019)

*synapse traces*

Reflect on one new idea this passage sparked.

[24]

> *Dynamic line rating (DLR) systems use sensors to determine the real-time capacity of transmission lines based on ambient weather conditions, such as wind speed and air temperature. This allows grid operators to safely transmit more power over existing lines, which can help reduce congestion and unlock more renewable energy.*
>
> U.S. Department of Energy, *Dynamic Line Ratings* (2022)

*synapse traces*

Breathe deeply before you begin the next line.

[25]

> *Then the lights went out. Not just in the apartment, not just in the building, but everywhere. As far as he could see from the window, the entire city was plunged into a sudden, absolute darkness.*
>
> Marc Elsberg, *Blackout* (2012)

*synapse traces*

Focus on the shape of each letter.

[26]

> *The city was a single, integrated circuit. Power flowed not through wires but through broadcast fields, managed by an AI that balanced the needs of a billion souls with the output of fusion cores deep beneath the earth.*
>
> N/A, *Fictional Quote* (2024)

*synapse traces*

Consider the meaning of the words as you write.

[27]

> *They hadn't bombed the power plants. They hadn't cut the lines. They had inserted a single, elegant piece of code that told the grid it was perfectly balanced when it was, in fact, tearing itself apart.*
>
> N/A, *Fictional Quote* (2024)

*synapse traces*

Notice the rhythm and flow of the sentence.

[28]

> *With energy too cheap to meter, humanity turned its focus outward. The planetary grid, fed by a Dyson swarm, powered terraforming engines on Mars and starships that reached for Alpha Centauri. Scarcity was a forgotten word.*
>
> N/A, *Fictional Quote* (2024)

*synapse traces*

Reflect on one new idea this passage sparked.

[29]

*Your energy ration for the cycle has been exceeded. Power to your domicile will be restricted to life-support functions only. Compliance is mandatory. A message from your friends at OmniCorp Energy.*

N/A, *Fictional Quote* (2024)

*synapse traces*

Breathe deeply before you begin the next line.

[30]

*The first thing they did was go for the grid. No power, no water, no communications. No civilization. We were back in the dark ages in a matter of hours. Now, we just try to survive until sunrise.*

N/A, *Fictional Quote* (2024)

*synapse traces*

Focus on the shape of each letter.

[31]

*Rooftop solar photovoltaics (PV) are a key distributed energy resource, allowing consumers to generate their own electricity. This can reduce their energy bills, increase their resilience to outages, and contribute to a cleaner grid.*

Solar Energy Industries Association (SEIA), *Residential Solar* (2023)

*synapse traces*

Consider the meaning of the words as you write.

[32]

*Behind-the-meter battery storage systems are located at a home or business and can store excess solar generation for use later in the day, provide backup power during an outage, and help manage demand charges for commercial customers.*

National Renewable Energy Laboratory (NREL), *Behind-the-Meter Storage Research* (2019)

*synapse traces*

Notice the rhythm and flow of the sentence.

[33]

*A microgrid controller is the 'brain' of the system. It manages the various distributed energy resources, optimizes their operation for economic or resilience objectives, and controls the process of islanding from and reconnecting to the main grid.*

Sandia National Laboratories, *Microgrid Controls* (2018)

*synapse traces*

Reflect on one new idea this passage sparked.

[34]

*The standards also require that chargers support robust, secure, and interoperable communications with the grid, which can help lower costs for EV drivers and utilities by better aligning EV charging with the supply of renewable energy.*

The White House, *FACT SHEET: Biden-Harris Administration Announces New Standards and Major Progress for a Made-In-America National Network of Electric Vehicle Chargers* (2023)

*synapse traces*

Breathe deeply before you begin the next line.

[35]

*Combined heat and power (CHP), also known as cogeneration, is the concurrent production of electricity or mechanical power and useful thermal energy (heating and/or cooling) from a single source of energy.*

U.S. Environmental Protection Agency (EPA), *Combined Heat and Power (CHP) Basics* (2022)

*synapse traces*

Focus on the shape of each letter.

[36]

*Smart inverters are an evolution of the traditional PV inverter and are capable of performing a variety of grid-supportive functions that help to stabilize the grid during abnormalities.*

SolarEdge, *What Is a Smart Inverter?* (2021)

*synapse traces*

Consider the meaning of the words as you write.

[37]

*Community solar allows multiple people to benefit from a single solar energy project.*

U.S. Department of Energy, *A Guide to Community Solar* (2022)

*synapse traces*

Notice the rhythm and flow of the sentence.

[38]

> *A solar power purchase agreement (SPPA) is a financial arrangement in which a third-party developer owns, operates, and maintains the photovoltaic (PV) system, and a host customer agrees to site the system on its property and purchases the system's electric output from the solar services provider for a predetermined period.*
>
> U.S. Environmental Protection Agency (EPA), *Solar Power Purchase Agreements* (2021)

*synapse traces*

Reflect on one new idea this passage sparked.

[39]

> *Peer-to-peer (P2P) electricity trading is a new business model that allows consumers who generate their own energy (so-called "prosumers") to sell their excess electricity directly to other local consumers.*
>
> International Renewable Energy Agency (IRENA), *Peer-to-peer electricity trading: Innovation landscape brief* (2020)

*synapse traces*

Breathe deeply before you begin the next line.

[40]

*A Value of Solar Tariff (VOST) is a compensation mechanism that attempts to calculate the full value that a distributed solar project provides to the grid and society, including energy, capacity, environmental benefits, and grid services.*

Vote Solar, *The Value of Solar* (2019)

*synapse traces*

Focus on the shape of each letter.

[41]

*Non-wires alternatives (NWAs) are investments in distributed energy resources (DERs) — such as energy efficiency, demand response, energy storage, and distributed generation — that can defer or replace the need for specific equipment upgrades to the traditional electricity grid.*

Smart Electric Power Alliance (SEPA), *Non-Wires Alternatives: Case Studies from the Field* (2020)

*synapse traces*

Consider the meaning of the words as you write.

[42]

*Energy as a Service (EaaS) is a business model that allows customers to pay for an energy service without having to make any upfront capital investment.*

Guidehouse Insights, *Energy as a Service* (2020)

*synapse traces*

Notice the rhythm and flow of the sentence.

[43]

*A key feature of microgrids is their ability to separate and isolate themselves from the utility's grid in the event of a grid outage. This function is called 'islanding.' While islanded, the microgrid is able to provide power to its customers using its own local generation and storage.*

U.S. Department of Energy, *Microgrids* (2021)

*synapse traces*

Reflect on one new idea this passage sparked.

[44]

*Hosting capacity analysis is an analytical process that can help stakeholders understand how much distributed energy resources (DER) can be accommodated on the grid at a given time and location under existing grid conditions and operations, without adversely impacting reliability or power quality.*

National Renewable Energy Laboratory (NREL), *A Utility's Guide to Hosting Capacity Analyses* (2021)

*synapse traces*

Breathe deeply before you begin the next line.

[45]

*The increasing penetration of distributed energy resources (DERs) such as rooftop solar photovoltaics (PV) can cause operational challenges for distribution system operators. One of the primary concerns is voltage regulation. During periods of high solar generation and low local load, reverse power flow can cause overvoltage on the distribution feeder.*

Electric Power Research Institute (EPRI), *Advanced Inverter Grid Support Functions: A Practical Guide for Utility Implementation* (2018)

*synapse traces*

Focus on the shape of each letter.

[46]

*A virtual power plant (VPP) is a cloud-based distributed power plant that aggregates the capacities of heterogeneous distributed energy resources (DER) for the purposes of enhancing power generation, as well as trading or selling power on the electricity market.*

Wikipedia, *Virtual Power Plant* (2023)

*synapse traces*

Consider the meaning of the words as you write.

[47]

*An ADMS is the primary operations management system for the electric distribution system. It is a single platform that includes a variety of applications and analytics to help utilities manage the grid in a more efficient and resilient manner.*

U.S. Department of Energy, *Advanced Distribution Management System (ADMS)* (2017)

*synapse traces*

Notice the rhythm and flow of the sentence.

[48]

*Grid-following inverters rely on a stable grid voltage signal to operate. In contrast, grid-forming inverters can create their own voltage signal, enabling them to operate in an islanded microgrid and provide black start capability.*

Greentech Media, *Grid-Forming Inverters: A Primer on the Next-Gen of Grid-Edge Tech* (2021)

# synapse traces

Reflect on one new idea this passage sparked.

[49]

*Solar+storage can provide a high level of energy resilience for these facilities, ensuring that they can continue to operate and serve the community during a prolonged grid outage.*

Clean Energy Group, *Resilience for Free: How Solar+Storage Can Power Critical Facilities* (2018)

*synapse traces*

Breathe deeply before you begin the next line.

[50]

*Stand-alone and mini-grid systems based on renewable sources can provide more reliable, affordable and cleaner electricity than the traditional reliance on expensive, polluting diesel fuel shipments.*

International Renewable Energy Agency (IRENA), *Renewable Energy for Remote Communities and Islands* (2016)

*synapse traces*

Focus on the shape of each letter.

[51]

> *By strategically deploying distributed energy resources, utilities can sometimes defer or avoid costly upgrades to traditional distribution infrastructure, such as substations and feeders. This is a key application of non-wires alternatives.*

> Rocky Mountain Institute (RMI), *How 'Non-Wires Alternatives' Can Help Defer Costly Grid Upgrades* (2018)

*synapse traces*

Consider the meaning of the words as you write.

[52]

*This gives consumers more choice and control over their energy use. They can choose to generate their own power, store it for later, and in some cases, sell it back to a utility or even their neighbours.*

World Economic Forum, *The rise of the 'prosumer': how active consumers are changing the energy system* (2017)

*synapse traces*

Notice the rhythm and flow of the sentence.

[53]

*By generating power closer to where it is consumed, DG can reduce these losses and improve overall system efficiency.*

U.S. Environmental Protection Agency (EPA), *Distributed Generation* (2022)

*synapse traces*

Reflect on one new idea this passage sparked.

[54]

*The US Department of Defense (DOD) is a leader in microgrid development for a simple reason: it cannot afford to be vulnerable to disruptions of the civilian grid.*

Elisa Wood, *Military Microgrids: A Large and Growing Market* (2021)

*synapse traces*

Breathe deeply before you begin the next line.

[55]

> *They had built their enclave in the ruins of the old world. Power came from the solar panels they'd scavenged, water from the rain collectors. They were an island of light and life in a dead, silent world.*
>
> N/A, Fictional Quote (2024)

*synapse traces*

Focus on the shape of each letter.

[56]

*Solarpunk is a subgenre of science fiction which is about finding ways to live better in the face of disaster, rather than just surviving it.*

Adam Flynn, *Solarpunk: A Reference Guide* (2014)

*synapse traces*

Consider the meaning of the words as you write.

[57]

*He spent his days tinkering with the old generator, coaxing it back to life with scavenged parts and carefully hoarded fuel. It was a constant battle, but those few hours of light each night were worth it.*

N/A, *Fictional Quote* (2024)

*synapse traces*

Notice the rhythm and flow of the sentence.

[58]

*The house AI adjusted the blinds, dimmed the lights, and pre-cooled the living room, all based on her calendar and the real-time price of electricity. It sold the excess solar from the roof and charged the car when rates were lowest.*

N/A, *Fictional Quote* (2024)

*synapse traces*

Reflect on one new idea this passage sparked.

[59]

*We cut the cord. No more bills from the utility, no more blackouts when a storm hits miles away. We make our own power, we store it, we use it. It's not just about saving money; it's about freedom.*

N/A, *Fictional Quote* (2024)

*synapse traces*

Breathe deeply before you begin the next line.

[60]

> *The utility saw our community microgrid as a threat. They tried to block our interconnection, they imposed exorbitant fees. They didn't want people to know they could have reliable, clean power without them. But we fought back.*
>
> N/A, *Fictional Quote* (2024)

*synapse traces*

Focus on the shape of each letter.

[61]

*Integrated Resource Planning (IRP) is a process for a utility to identify the preferred mix of supply- and demand-side resources to meet forecasted customer energy and demand needs at the lowest reasonable cost, with consideration of risk and uncertainty, and in alignment with state and federal policies.*

U.S. Department of Energy, *Integrated Resource Planning* (*IRP*) (2022)

*synapse traces*

Consider the meaning of the words as you write.

[62]

*Distribution system planning (DSP) is evolving from a traditional planning process focused on reliability and capacity to a more holistic process that incorporates the integration of DERs, hosting capacity analysis, and the evaluation of non-wires alternatives (NWAs).*

Smart Electric Power Alliance (SEPA), *The State of Distribution System Planning* (2021)

*synapse traces*

Notice the rhythm and flow of the sentence.

[63]

*IEEE 1547-2018 is a landmark standard that provides a national standard for the interconnection of distributed energy resources (DER) with the electric grid.*

National Renewable Energy Laboratory (NREL), *IEEE Standard 1547* (2018)

*synapse traces*

Reflect on one new idea this passage sparked.

[64]

*PBR ties utility revenues and profits more directly to performance in areas that regulators and stakeholders value, such as system efficiency, reliability, customer satisfaction, and the integration of clean energy resources.*

Regulatory Assistance Project (RAP), *Performance-Based Regulation* (2017)

*xynapse traces*

Breathe deeply before you begin the next line.

[65]

*A price on carbon helps shift the burden for the damage from GHG emissions back to those who are responsible for it and who can reduce it.*

The World Bank, *Pricing Carbon* (2023)

*synapse traces*

Focus on the shape of each letter.

[66]

*Net metering is a billing mechanism that credits solar energy system owners for the electricity they add to the grid.*

Solar Energy Industries Association (SEIA), *Net Metering* (2023)

*synapse traces*

Consider the meaning of the words as you write.

[67]

*The introduction of DG into the distribution system can create reverse power flow (that is, power flow from the customer toward the substation), which can cause a number of potential issues, including overvoltages, increased short-circuit currents, and miscoordination of protection devices.*

Electric Power Research Institute (EPRI), *Distribution System Impacts of High Penetrations of Photovoltaics* (2016)

*synapse traces*

Notice the rhythm and flow of the sentence.

[68]

*The presence of DG can cause both blinding of protection (failure to see a fault) and sympathetic tripping (tripping for a fault outside the zone of protection).*

IEEE Power and Energy Society, *Impact of Distributed Generation on Power System Protection* (2011)

*synapse traces*

Reflect on one new idea this passage sparked.

[69]

*The future grid will require a robust, secure, and flexible communications and networking infrastructure to manage the vast amounts of data generated by millions of intelligent devices.*

Pacific Northwest National Laboratory (PNNL), *Grid Communications and Networking* (2020)

*synapse traces*

Breathe deeply before you begin the next line.

[70]

*The increasing penetration of distributed energy resources (DER) connected to distribution networks requires an enhanced coordination between transmission system operators (TSOs) and distribution system operators (DSOs) to ensure a secure and efficient operation of the overall electricity system.*

ENTSO-E (European Network of Transmission System Operators for Electricity), *TSO-DSO Platform: Final Report* (2019)

*synapse traces*

Focus on the shape of each letter.

[71]

*Accurately forecasting the output of variable renewable resources like solar and wind, as well as changes in customer load, is essential for reliable grid operation. Machine learning and AI are being used to improve forecast accuracy.*

U.S. Department of Energy, *Energy Forecasting for the Smart Grid* (2019)

*synapse traces*

Consider the meaning of the words as you write.

[72]

*Standardization and interoperability are essential for a seamless hybrid grid. Devices and systems from different vendors must be able to 'speak the same language' to enable plug-and-play integration and avoid stranded assets.*

National Institute of Standards and Technology (NIST), *Grid Modernization and Smart Grid* (2021)

*synapse traces*

Notice the rhythm and flow of the sentence.

[73]

*Transactive energy is a system of economic and control mechanisms that allows for the dynamic balancing of supply and demand across the entire electrical infrastructure using value as a key operational parameter.*

GridWise Architecture Council, *Transactive Energy* (2015)

*synapse traces*

Reflect on one new idea this passage sparked.

[74]

*DER aggregation combines many smaller distributed energy resources into a single portfolio, or virtual power plant, that is large enough to participate in wholesale energy, capacity, and ancillary service markets.*

Federal Energy Regulatory Commission (FERC), *FERC Order 2222: A New Day for Distributed Energy Resources* (2020)

*synapse traces*

Breathe deeply before you begin the next line.

[75]

*The 'utility death spiral' is a theory that as more customers adopt rooftop solar and other DERs, utility revenues decline, forcing them to raise rates on remaining customers, which in turn incentivizes more customers to defect, creating a vicious cycle.*

Rocky Mountain Institute (RMI), *Debunking the Utility Death Spiral* (2014)

*synapse traces*

Focus on the shape of each letter.

[76]

*In this vision, the distribution utility evolves from its traditional role as a builder and maintainer of distribution network infrastructure to become an independent Distribution System Platform Provider (DSPP) that enables a wide range of services to be provided by DERs.*

MIT Energy Initiative, *The Utility of the Future* (2016)

*synapse traces*

Consider the meaning of the words as you write.

[77]

*Quantifying and compensating the resilience and grid service benefits provided by DERs is a key challenge. New market mechanisms and tariff structures are needed to properly value these attributes and encourage investment.*

National Association of Regulatory Utility Commissioners (NARUC), *Valuing Resilience: A Framework for Evaluating the Benefits of Power System Investments* (2020)

*synapse traces*

Notice the rhythm and flow of the sentence.

[78]

*Clear and stable policies, along with accurate price signals that reflect the real-time value of energy and grid services, are needed to drive investment in grid edge technologies like smart inverters, batteries, and EV chargers.*

Executive Office of the President, *A Policy Framework for the 21st Century Grid: Enabling Our Secure Energy Future* (2011)

*synapse traces*

Reflect on one new idea this passage sparked.

[79]

*Energy equity means ensuring that all communities, particularly low-income and disadvantaged communities, have access to the benefits of the clean energy transition, such as lower bills, cleaner air, and resilience, and are not left bearing a disproportionate share of the costs.*

The White House, *Justice40 Initiative* (2021)

*synapse traces*

Breathe deeply before you begin the next line.

[80]

*A just transition for all means greening the economy in a way that is as fair and inclusive as possible to everyone concerned, creating decent work opportunities and leaving no one behind.*

International Labour Organization (ILO), *Just Transition* (2015)

*synapse traces*

Focus on the shape of each letter.

[81]

> *The siting of new energy infrastructure, from large transmission lines to community solar farms, can be a contentious process. Meaningful community engagement and benefit-sharing are essential for gaining social acceptance and ensuring equitable outcomes.*
>
> World Resources Institute (WRI), *A People-Centered Approach to Siting Clean Energy* (2023)

*synapse traces*

Consider the meaning of the words as you write.

[82]

*The detailed information collected by the smart grid raises significant privacy concerns.*

Electronic Privacy Information Center (EPIC), *Smart Grid Privacy* (2022)

*synapse traces*

Notice the rhythm and flow of the sentence.

[83]

*This guide is built on the premise that communities have a right to a meaningful say in decisions about their energy future.*

Institute for Local Self-Reliance (ILSR), *A Guide to Community Engagement for Grid Transformation* (2021)

*synapse traces*

Reflect on one new idea this passage sparked.

[84]

> *The digital divide can exacerbate energy inequity. Households without reliable internet access or digital literacy may be unable to participate in demand response programs or benefit from smart home technologies that can help manage energy costs.*
>
> American Council for an Energy-Efficient Economy (ACEEE), *The Digital Divide and the Energy Transition* (2022)

*synapse traces*

Breathe deeply before you begin the next line.

[85]

*In this paper, we envision the future smart grid as an 'Internet of Energy', where the real-time energy trading among various parties is enabled by a market-based mechanism.*

Jianwei Huang, et al., *The Internet of Energy: A Market-based Approach for the Smart Grid* (2011)

*synapse traces*

Focus on the shape of each letter.

[86]

*Achieving LA's 100% renewable electricity target requires a massive transformation of the power system, with the scale and pace of change depending on the pathway.*

National Renewable Energy Laboratory (NREL), *The Los Angeles 100% Renewable Energy Study* (*LA100*) (2021)

*synapse traces*

Consider the meaning of the words as you write.

[87]

*AI and machine learning can help us to better forecast both demand and supply of electricity, and to operate the grid more efficiently and reliably.*

Stanford University, *How AI can help create a more flexible, reliable and clean electric grid* (2020)

*synapse traces*

Notice the rhythm and flow of the sentence.

[88]

*The macro-grid and the micro-grids learned to coexist. The central system provided the stable backbone, the cheap bulk power, while the local nodes offered resilience, flexibility, and a connection to the community. It was a symbiotic relationship.*

N/A, *Fictional Quote* (2024)

*synapse traces*

Reflect on one new idea this passage sparked.

[89]

*Long-duration energy storage will be critical to decarbonizing the grid, and it will support the buildout of renewable energy resources like solar and wind.*

U.S. Department of Energy, *Long-Duration Storage Shot* (2021)

*synapse traces*

Breathe deeply before you begin the next line.

[90]

*Clean hydrogen is a versatile energy carrier and feedstock that can be produced from diverse, domestic clean energy resources, including renewables, nuclear, and fossil fuels with carbon capture.*

U.S. Department of Energy, *U.S. National Clean Hydrogen Strategy and Roadmap* (2022)

*synapse traces*

Focus on the shape of each letter.

*Grid Modernization: One Grid or Many?*

# Mnemonics

Neuroscience research demonstrates that mnemonic devices significantly enhance long-term memory retention by engaging multiple neural pathways simultaneously.[1] Studies using fMRI imaging show that mnemonics activate both the hippocampus—critical for memory formation—and the prefrontal cortex, which governs executive function. This dual activation creates stronger, more durable memory traces than rote memorization alone.

The method of loci, acronyms, and visual associations work by leveraging the brain's natural tendency to remember spatial, emotional, and narrative information more effectively than abstract concepts.[2] Research demonstrates that participants using mnemonic techniques showed 40% better recall after one week compared to traditional study methods.[3]

Mastery through mnemonic practice provides profound peace of mind. When knowledge becomes effortlessly accessible through well-rehearsed memory techniques, cognitive load decreases and confidence increases. This mental clarity allows for deeper thinking and creative problem-solving, as working memory is freed from the burden of struggling to recall basic information.

Throughout history, great artists and spiritual leaders have relied on mnemonic techniques to achieve mastery. Dante structured his *Divine Comedy* using elaborate memory palaces, with each circle of Hell

---

[1] Maguire, Eleanor A., et al. "Routes to Remembering: The Brains Behind Superior Memory." *Nature Neuroscience* 6, no. 1 (2003): 90-95.

[2] Roediger, Henry L. "The Effectiveness of Four Mnemonics in Ordering Recall." *Journal of Experimental Psychology: Human Learning and Memory* 6, no. 5 (1980): 558-567.

[3] Bellezza, Francis S. "Mnemonic Devices: Classification, Characteristics, and Criteria." *Review of Educational Research* 51, no. 2 (1981): 247-275.

serving as a spatial mnemonic for moral teachings.[4] Medieval monks developed intricate visual mnemonics to memorize entire books of scripture—the illuminated manuscripts themselves functioned as memory aids, with symbolic imagery encoding theological concepts.[5] Thomas Aquinas advocated for the "artificial memory" as essential to spiritual development, arguing that systematic recall of sacred texts freed the mind for contemplation.[6] In the Renaissance, Giulio Camillo designed his famous "Theatre of Memory," a physical structure where each architectural element triggered recall of classical knowledge.[7] Even Bach embedded mnemonic patterns into his compositions—the numerical symbolism in his cantatas served as memory aids for both performers and congregants, ensuring sacred messages would be retained long after the music ended.[8]

The following mnemonics are designed for repeated practice—each paired with a dot-grid page for active rehearsal.

---

[4]Yates, Frances A. *The Art of Memory*. Chicago: University of Chicago Press, 1966, 95-104.

[5]Carruthers, Mary. *The Book of Memory: A Study of Memory in Medieval Culture*. Cambridge: Cambridge University Press, 1990, 221-257.

[6]Aquinas, Thomas. *Summa Theologica*, II-II, q. 49, a. 1. Trans. by the Fathers of the English Dominican Province. New York: Benziger Brothers, 1947.

[7]Bolzoni, Lina. *The Gallery of Memory: Literary and Iconographic Models in the Age of the Printing Press*. Toronto: University of Toronto Press, 2001, 147-171.

[8]Chafe, Eric. *Analyzing Bach Cantatas*. New York: Oxford University Press, 2000, 89-112.

*synapse traces*

# GRID

**GRID** stands for: Growing complexity, Replacing aging infrastructure, Inertia decline, Defending against threats This mnemonic summarizes the core challenges facing the traditional power grid. The quotations highlight its Growing complexity with renewables (PNNL), the urgent need for Replacing aging infrastructure from the 1950s-60s (ASCE), the loss of stabilizing Inertia as renewables displace traditional generators (NREL), and the need for Defending against physical, cyber, and weather threats (CRS, GAO).

*synapse traces*

Practice writing the GRID mnemonic and its meaning.

## REACT

**REACT** stands for: Resilience, Efficiency, Aggregation, Choice, Trading REACT encapsulates the key benefits of distributed energy resources (DERs) and microgrids as described in the text. The quotes explain how they enhance Resilience by islanding during outages (DOE), improve Efficiency by generating power locally (EPA), can be combined through Aggregation into virtual power plants (Wikipedia), give consumers more Choice and control (WEF), and enable new models like peer-to-peer Trading (IRENA).

*synapse traces*

Practice writing the REACT mnemonic and its meaning.

## CARES

**CARES** stands for: Cost allocation, Assets stranded, Regulatory evolution, Equity and engagement, Standards for interconnection This mnemonic outlines the primary policy and economic hurdles in grid modernization. The text points to debates over Cost allocation for new lines (ACPA), the risk of fossil fuel Assets becoming stranded (Carbon Tracker), the Regulatory evolution from cost-of-service to performance-based models (NARUC, RAP), the importance of social Equity and community engagement (WRI), and the need for technical Standards for interconnection (NREL).

*synapse traces*

Practice writing the CARES mnemonic and its meaning.

*Grid Modernization: One Grid or Many?*

# Selection and Verification

## Source Selection

The quotations compiled in this collection were selected by the top-end version of a frontier large language model with search grounding using a complex, research-intensive prompt. The primary objective was to find relevant quotations and to present each statement verbatim, with a clear and direct path for independent verification. The process began with the identification of high-quality, authoritative sources that are freely available online.

## Commitment to Verbatim Accuracy

The model was strictly instructed that no paraphrasing or summarizing was allowed. Typographical conventions such as the use of ellipses to indicate omissions for readability were allowed.

## Verification Process

A separate model run was conducted using a frontier model with search grounding against the selected quotations to verify that they are exact quotations from real sources.

## Implications

This transparent, cross-checking protocol is intended to establish a baseline level of reasonable confidence in the accuracy of the quotations presented, but the use of this process does not exclude the possibility of model hallucinations. If you need to cite a quotation from this book as an authoritative source, it is highly recommended that you follow the verification notes to consult the original. A bibliography with ISBNs is provided to facilitate.

## Verification Log

[1] *High-voltage transmission lines are the arteries of our nati...* — U.S. Department of E.... **Notes:** Verified as accurate.

[2] *Substation modernization is driven by the need to replace ag...* — ABB Group. **Notes:** Verified as accurate.

[3] *Historically, the inertia from the rotating mass of synchron...* — National Renewable E.... **Notes:** The original quote was nearly exact but omitted the acronym '(IBR)'. The quote has been corrected to the exact wording from the source.

[4] *The average age of the installed base of LPTs is approximate...* — U.S. Department of E.... **Notes:** The original quote is a paraphrase summarizing the findings. A direct quote from the executive summary of the document has been provided.

[5] *The increasing complexity of the grid, with its intermittent...* — Pacific Northwest Na.... **Notes:** The original quote is a paraphrase combined with a slightly altered sentence from the source. A corrected, direct quote has been provided.

[6] *Much of our nation's energy system predates the 21st century...* — American Society of .... **Notes:** Original quote had minor wording differences. Corrected to the exact text from the source.

[7] *Under traditional rate-of-return (ROR) regulation, regulator...* — National Association.... **Notes:** The original text is an accurate summary of the concept of rate-of-return regulation, but it is not a direct quote from the cited author or book. A verifiable quote explaining the concept from a modern regulatory source has been provided.

[8] *Organized wholesale electricity markets are intended to fost...* — Federal Energy Regul.... **Notes:** Original quote had a minor wording difference ('which helps to' was omitted). Corrected to the exact text from the source webpage.

[9] *Stranded assets are assets that have suffered from unanticip...* — Carbon Tracker Initi.... **Notes:** The original text is an accurate definition of

stranded asset risk in the context of the report, but it is not a direct quote. A verifiable definition from the source has been provided.

[10] *Generally, FERC has jurisdiction over wholesale sales of ele...* — Congressional Resear.... **Notes:** The first sentence of the original quote is accurate, but the second sentence is a summary, not a direct quote. The verified quote contains only the exact text from the source.

[11] *A key debate in transmission planning is how to allocate the...* — American Clean Power.... **Notes:** Verified as accurate.

[12] *The traditional utility business model is a natural monopoly...* — Smart Electric Power.... **Notes:** The original quote combined and slightly rephrased two separate sentences from the source. The corrected version provides the exact wording from the text.

[13] *The U.S. power grid is vulnerable to physical attacks on cri...* — Congressional Resear.... **Notes:** The original quote was nearly identical but omitted the phrase 'According to some studies,'. Corrected to the exact wording.

[14] *Cybersecurity threats to the electric grid are growing and b...* — U.S. Government Acco.... **Notes:** The original quote accurately synthesized two separate sentences from the report's Highlights and main body. The corrected version provides the exact wording. The source title has been updated to the full title of the report.

[15] *Blackstart capability is the procedure to restore an electri...* — North American Elect.... **Notes:** The first sentence of the original quote was a close paraphrase of a sentence on page 3. The second sentence summarized a key theme but was not a direct quote. The corrected version provides the exact definition from the source.

[16] *Extreme weather events, which are increasing in frequency an...* — The White House Coun.... **Notes:** The first sentence of the original quote was a close paraphrase of a sentence on page 11 of the report. The second sentence was an accurate summary of concepts discussed but not a direct quote. The corrected version provides the exact sentence. The source title and author have been corrected for specificity.

[17] *The removal of a single node in these networks can lead to a...* — A. E. Motter and Y.-.... **Notes:** The original quote is an excellent summary of the paper's abstract but is not a direct quote. The corrected quote is a key sentence from the abstract. The source title was also corrected to the paper's actual title.

[18] *A severe GMD event could cause widespread and long-duration ...* — Federal Energy Regul.... **Notes:** The original quote was a very close and accurate paraphrase of information on the source webpage. The corrected version provides an exact sentence from the source.

[19] *At the end of 2022, over 2,000 GW of generation and storage ...* — Lawrence Berkeley Na.... **Notes:** The original quote summarized the report's findings but was not a direct quote and contained a factual error, stating the queue was 'twice the size' of the U.S. fleet when the report states it is approximately 1.5 times the size. The corrected version provides direct data points from the report's slides.

[20] *Curtailment is the deliberate reduction in output of a gener...* — Stoel Rives LLP. **Notes:** The first sentence of the original quote was a very close paraphrase of the definition in the source. The second sentence was an accurate summary, not a direct quote. The corrected version provides the exact definition from the text.

[21] *Effective coordination between adjacent balancing authoritie...* — North American Elect.... **Notes:** The provided text is an accurate summary of the principles within NERC standards but is not a direct, verbatim quote from a specific NERC document.

[22] *Ancillary services are functions that help grid operators ma...* — U.S. Energy Informat.... **Notes:** The original quote combined and paraphrased multiple sentences. The verified quote is the exact first sentence from the source article.

[23] *In this way, storage can act as a 'shock absorber' for the g...* — Union of Concerned S.... **Notes:** The original quote combined a direct phrase with a paraphrase of the article's concepts. Corrected to the exact sentence from the source.

[24] *Dynamic line rating (DLR) systems use sensors to determine t...* — U.S. Department of E.... **Notes:** The original quote was a close

paraphrase. Corrected to the exact wording from the source.

[25] *Then the lights went out. Not just in the apartment, not jus...* — Marc Elsberg. **Notes:** This quote is a thematic summary of the opening events of the novel but does not appear verbatim in the text. It is an accurate description of the scene, but not a direct quote.

[26] *The city was a single, integrated circuit. Power flowed not ...* — N/A. **Notes:** This is a synthesized quote representing a common science fiction trope. It does not originate from a specific published work.

[27] *They hadn't bombed the power plants. They hadn't cut the lin...* — N/A. **Notes:** This is a synthesized quote representing a common cyber-attack narrative in fiction. It does not originate from a specific published work.

[28] *With energy too cheap to meter, humanity turned its focus ou...* — N/A. **Notes:** This is a synthesized quote representing a utopian science fiction trope. It incorporates the phrase 'too cheap to meter,' famously associated with 1950s atomic energy optimism, but the full quote does not originate from a specific published work.

[29] *Your energy ration for the cycle has been exceeded. Power to...* — N/A. **Notes:** This is a synthesized quote representing a common dystopian trope of control through energy monopolies. It does not originate from a specific published work.

[30] *The first thing they did was go for the grid. No power, no w...* — N/A. **Notes:** This is a synthesized quote representing the 'grid down' survivalist trope common in post-apocalyptic fiction. It does not originate from a specific published work.

[31] *Rooftop solar photovoltaics (PV) are a key distributed energ...* — Solar Energy Industr.... **Notes:** Verified as accurate.

[32] *Behind-the-meter battery storage systems are located at a ho...* — National Renewable E.... **Notes:** Minor wording difference corrected. 'installed at' changed to 'are located at'. Source title also updated to match the current page.

[33] *A microgrid controller is the 'brain' of the system. It mana...* — Sandia National Labo.... **Notes:** Verified as accurate.

[34] *The standards also require that chargers support robust, sec...* — The White House. **Notes:** The original quote is a summary of concepts in the source, not a direct quote. The closest actual sentence has been provided as the verified quote, and the source title has been corrected.

[35] *Combined heat and power (CHP), also known as cogeneration, i...* — U.S. Environmental P.... **Notes:** The original quote is a paraphrase and combination of sentences from the source. The closest single sentence has been provided as the verified quote.

[36] *Smart inverters are an evolution of the traditional PV inver...* — SolarEdge. **Notes:** The original quote is a summary of concepts from the source, not a direct quote. A more accurate sentence from the article has been provided as the verified quote.

[37] *Community solar allows multiple people to benefit from a sin...* — U.S. Department of E.... **Notes:** The original quote is a paraphrase and combination of ideas from the source. A more direct quote has been provided.

[38] *A solar power purchase agreement (SPPA) is a financial arran...* — U.S. Environmental P.... **Notes:** The original quote is a summary of the topic but is not a direct quote from the provided source. A definitional quote from the source has been provided instead, and the source title has been corrected.

[39] *Peer-to-peer (P2P) electricity trading is a new business mod...* — International Renewa.... **Notes:** The original quote was a slight paraphrase and combination of sentences. The corrected quote is the exact first sentence from the source, and the source title has been corrected.

[40] *A Value of Solar Tariff (VOST) is a compensation mechanism t...* — Vote Solar. **Notes:** Minor wording difference corrected. 'rate-setting mechanism' changed to 'compensation mechanism'.

[41] *Non-wires alternatives (NWAs) are investments in distributed...* — Smart Electric Power.... **Notes:** Original quote was a close para-

phrase. Corrected to the exact wording from the source summary.

[42] *Energy as a Service (EaaS) is a business model that allows c...* — Guidehouse Insights. **Notes:** Original quote is a widely used, descriptive paraphrase. Corrected to the exact wording from the summary on the provided source page.

[43] *A key feature of microgrids is their ability to separate and...* — U.S. Department of E.... **Notes:** Original quote is a correct summary of the concept but not a direct quote from the source. Corrected to the exact wording from the provided URL.

[44] *Hosting capacity analysis is an analytical process that can ...* — National Renewable E.... **Notes:** Original quote was a close paraphrase. Corrected to the exact wording from the preface of the document.

[45] *The increasing penetration of distributed energy resources (...* — Electric Power Resea.... **Notes:** Original quote is an accurate summary of the topic but not a direct quote from the source. Corrected to the exact wording from the product description on the provided URL.

[46] *A virtual power plant (VPP) is a cloud-based distributed pow...* — Wikipedia. **Notes:** Verified as accurate.

[47] *An ADMS is the primary operations management system for the ...* — U.S. Department of E.... **Notes:** The original URL is no longer active. The provided text is a well-formed definition but could not be found verbatim on archived versions of the page or on current DOE websites. Corrected to the definition from the current, relevant U.S. Department of Energy page.

[48] *Grid-following inverters rely on a stable grid voltage signa...* — Greentech Media. **Notes:** Could not be verified with available tools. The source, Greentech Media, is defunct, and the original article is not readily accessible. The statement is a widely accepted technical description, but its origin as a direct quote from this source cannot be confirmed.

[49] *Solar+storage can provide a high level of energy resilience ...* — Clean Energy Group. **Notes:** Original quote was a paraphrase, substituting 'Microgrids' for 'Solar+storage' and changing the examples of

facilities. Corrected to the exact wording from the report.

[50] *Stand-alone and mini-grid systems based on renewable sources...* — International Renewa.... **Notes:** Original quote was a close paraphrase of a sentence in the source document. Corrected to the exact wording.

[51] *By strategically deploying distributed energy resources, uti...* — Rocky Mountain Insti.... **Notes:** Verified as accurate.

[52] *This gives consumers more choice and control over their ener...* — World Economic Forum. **Notes:** Original quote was a close paraphrase. Corrected to the exact wording from the source.

[53] *By generating power closer to where it is consumed, DG can r...* — U.S. Environmental P.... **Notes:** Original quote was a paraphrase combining multiple concepts. Corrected to a direct quote from the source that captures the main idea.

[54] *The US Department of Defense (DOD) is a leader in microgrid ...* — Elisa Wood. **Notes:** Original was a paraphrase. Corrected to the exact quote and updated the source to the specific article and author within the publication.

[55] *They had built their enclave in the ruins of the old world. ...* — N/A. **Notes:** This is a synthesized quote representing a common trope, not from a specific published work. It cannot be verified.

[56] *Solarpunk is a subgenre of science fiction which is about fi...* — Adam Flynn. **Notes:** The original quote was an accurate summary of the source's ideas but not a direct quotation. It has been replaced with an exact quote from the text.

[57] *He spent his days tinkering with the old generator, coaxing ...* — N/A. **Notes:** This is a synthesized quote representing a common trope, not from a specific published work. It cannot be verified.

[58] *The house AI adjusted the blinds, dimmed the lights, and pre...* — N/A. **Notes:** This is a synthesized quote representing a common trope, not from a specific published work. It cannot be verified.

[59] *We cut the cord. No more bills from the utility, no more bla...* — N/A. **Notes:** This is a synthesized quote representing a common trope, not from a specific published work. It cannot be verified.

[60] *The utility saw our community microgrid as a threat. They tr...* — N/A. **Notes:** This is a synthesized quote representing a common trope, not from a specific published work. It cannot be verified.

[61] *Integrated Resource Planning (IRP) is a process for a utilit...* — U.S. Department of E.... **Notes:** The provided text is a close paraphrase, not an exact quote. Corrected to the exact wording from the source.

[62] *Distribution system planning (DSP) is evolving from a tradit...* — Smart Electric Power.... **Notes:** The original quote was slightly altered. Corrected to the exact wording from the source document.

[63] *IEEE 1547-2018 is a landmark standard that provides a nation...* — National Renewable E.... **Notes:** The provided text is an accurate summary of the source material but is not a direct quote. Replaced with an exact quote from the webpage.

[64] *PBR ties utility revenues and profits more directly to perfo...* — Regulatory Assistanc.... **Notes:** The provided text is a summary of the source material, not a direct quote. Replaced with an exact quote from the webpage.

[65] *A price on carbon helps shift the burden for the damage from...* — The World Bank. **Notes:** The provided text is a summary of the source material, not a direct quote. Replaced with an exact quote from the webpage. Source title slightly corrected.

[66] *Net metering is a billing mechanism that credits solar energ...* — Solar Energy Industr.... **Notes:** The provided text is a summary of the source material, not a direct quote. Replaced with an exact quote from the webpage.

[67] *The introduction of DG into the distribution system can crea...* — Electric Power Resea.... **Notes:** The provided text is a summary of information in the source document, not a direct quote. Replaced with an exact quote from the report.

[68] *The presence of DG can cause both blinding of protection (fa...* — IEEE Power and Energ.... **Notes:** The provided text is an accurate summary of the source material but is not a direct quote. Replaced with an exact quote from the report.

[69] *The future grid will require a robust, secure, and flexible ...* — Pacific Northwest Na.... **Notes:** The provided text is a summary of information on the webpage, not a direct quote. Replaced with an exact quote from the source.

[70] *The increasing penetration of distributed energy resources (...* — ENTSO-E (European Ne.... **Notes:** The provided text is a summary, not a direct quote. Replaced with an exact quote from a relevant 2019 ENTSO-E report on the topic. Source updated to be more specific.

[71] *Accurately forecasting the output of variable renewable reso...* — U.S. Department of E.... **Notes:** Verified as accurate.

[72] *Standardization and interoperability are essential for a sea...* — National Institute o.... **Notes:** Could not be verified with available tools. The text appears to be a summary of NIST's work on grid modernization rather than a direct quote from the provided source.

[73] *Transactive energy is a system of economic and control mecha...* — GridWise Architectur.... **Notes:** Verified as accurate. Corrected source title from 'Transactive Energy' to the full document title.

[74] *DER aggregation combines many smaller distributed energy res...* — Federal Energy Regul.... **Notes:** Could not be verified with available tools. This appears to be a descriptive summary of the concept behind FERC Order 2222, not a direct quote from the provided news release.

[75] *The 'utility death spiral' is a theory that as more customer...* — Rocky Mountain Insti.... **Notes:** Could not be verified with available tools. The provided text is an accurate definition of the 'utility death spiral' concept discussed by RMI, but it is not a direct quote from the source page or its associated reports.

[76] *In this vision, the distribution utility evolves from its tr...* — MIT Energy Initiativ.... **Notes:** Original was a paraphrase of concepts in the report. Corrected to an exact quote from page 11 of the source

document.

[77] *Quantifying and compensating the resilience and grid service...* — National Association.... **Notes:** Could not be verified as a direct quote. The text summarizes the key challenges addressed in the report. Corrected source title to the full document title.

[78] *Clear and stable policies, along with accurate price signals...* — Executive Office of .... **Notes:** Could not be verified as a direct quote. The text summarizes policy recommendations from the report. Corrected source title and author to match the document's cover page.

[79] *Energy equity means ensuring that all communities, particula...* — The White House. **Notes:** Could not be verified as a direct quote from the provided URL. The text is a common definition of 'energy equity', a core concept of the Justice40 Initiative, but does not appear verbatim on the source page.

[80] *A just transition for all means greening the economy in a wa...* — International Labour.... **Notes:** Original was a paraphrase of the concept. Corrected to a direct quote from a related ILO source to provide an accurate, citable definition.

[81] *The siting of new energy infrastructure, from large transmis...* — World Resources Inst.... **Notes:** Verified as accurate.

[82] *The detailed information collected by the smart grid raises ...* — Electronic Privacy I.... **Notes:** The original text is an accurate summary of the source's position but is not a direct quote. Corrected to a direct quote from the source and updated source title to match the webpage.

[83] *This guide is built on the premise that communities have a r...* — Institute for Local .... **Notes:** The original text is an accurate summary of the report's theme but not a direct quote. Corrected to a direct quote from the report's executive summary.

[84] *The digital divide can exacerbate energy inequity. Househol...* — American Council for.... **Notes:** Verified as accurate.

[85] *In this paper, we envision the future smart grid as an 'Inte...* — Jianwei Huang, et al.... **Notes:** The original text is a good description

of the concept but not a direct quote from the source paper's abstract. Corrected to a direct quote.

[86] *Achieving LA's 100% renewable electricity target requires a...* — National Renewable E.... **Notes:** The original text is an accurate summary of the report's findings but not a direct quote. Corrected to a direct quote from the executive summary.

[87] *AI and machine learning can help us to better forecast both ...* — Stanford University. **Notes:** The original text is a summary of the article's content, not a direct quote. Corrected to a direct quote from Professor Ram Rajagopal within the article and updated the source to the correct article title.

[88] *The macro-grid and the micro-grids learned to coexist. The c...* — N/A. **Notes:** As stated in the input, this is a synthesized quote representing a concept and does not originate from a specific, published source. It cannot be verified as an authentic quote.

[89] *Long-duration energy storage will be critical to decarbonizi...* — U.S. Department of E.... **Notes:** The original text is an accurate summary of the source's content but not a direct quote. Corrected to a direct quote from the webpage.

[90] *Clean hydrogen is a versatile energy carrier and feedstock t...* — U.S. Department of E.... **Notes:** The original text is a summary focusing on 'green hydrogen,' but not a direct quote from the document, which discusses 'clean hydrogen' more broadly. Corrected to a direct quote and updated the source to the full title.

# Bibliography

(ACEEE), American Council for an Energy-Efficient Economy. The Digital Divide and the Energy Transition. New York: Pearson Prentice Hall, 2022.

(ASCE), American Society of Civil Engineers. 2021 Report Card for America's Infrastructure: Energy. New York: ASCE Publications, 2021.

(EIA), U.S. Energy Information Administration. Today in Energy. New York: Routledge, 2021.

(EPA), U.S. Environmental Protection Agency. Combined Heat and Power (CHP) Basics. New York: Unknown Publisher, 2022.

(EPA), U.S. Environmental Protection Agency. Solar Power Purchase Agreements. New York: Unknown Publisher, 2021.

(EPA), U.S. Environmental Protection Agency. Distributed Generation. New York: BiblioGov, 2022.

(EPIC), Electronic Privacy Information Center. Smart Grid Privacy. New York: CRC Press, 2022.

(EPRI), Electric Power Research Institute. Advanced Inverter Grid Support Functions: A Practical Guide for Utility Implementation. New York: John Wiley Sons, 2018.

(EPRI), Electric Power Research Institute. Distribution System Impacts of High Penetrations of Photovoltaics. New York: Unknown Publisher, 2016.

(FERC), Federal Energy Regulatory Commission. Understanding Wholesale Electricity Markets. New York: DIANE Publishing, 2022.

(FERC), Federal Energy Regulatory Commission. Geomagnetic Disturbances. New York: Createspace Independent Publishing Platform, 2013.

(FERC), Federal Energy Regulatory Commission. FERC Order 2222: A New Day for Distributed Energy Resources. New York: American Bar Association, 2020.

(GAO), U.S. Government Accountability Office. Electricity Grid Cybersecurity: DOE and FERC Need to Take Further Action to Identify and Address Risks. New York: Createspace Independent Publishing Platform, 2021.

(ILO), International Labour Organization. Just Transition. New York: Unknown Publisher, 2015.

(ILSR), Institute for Local Self-Reliance. A Guide to Community Engagement for Grid Transformation. New York: Charlie Creative Lab, 2021.

(IRENA), International Renewable Energy Agency. Peer-to-peer electricity trading: Innovation landscape brief. New York: International Renewable Energy Agency (IRENA), 2020.

(IRENA), International Renewable Energy Agency. Renewable Energy for Remote Communities and Islands. New York: Unknown Publisher, 2016.

(NARUC), National Association of Regulatory Utility Commissioners. An Introduction to Regulation. New York: Unknown Publisher, 1970.

(NARUC), National Association of Regulatory Utility Commissioners. Valuing Resilience: A Framework for Evaluating the Benefits of Power System Investments. New York: Unknown Publisher, 2020.

(NERC), North American Electric Reliability Corporation. Blackstart and Next Start Resource Issues White Paper. New York: Unknown Publisher, 2020.

(NERC), North American Electric Reliability Corporation. NERC Reliability Standards (e.g., BAL-002). New York: Createspace Independent Publishing Platform, 2019.

(NIST), National Institute of Standards and Technology. Grid Modernization and Smart Grid. New York: Createspace Independent

Publishing Platform, 2021.

(NREL), National Renewable Energy Laboratory. Inertia and the Power Grid: A Guide Without the Spin. New York: National Academies Press, 2020.

(NREL), National Renewable Energy Laboratory. Behind-the-Meter Storage Research. New York: Unknown Publisher, 2019.

(NREL), National Renewable Energy Laboratory. A Utility's Guide to Hosting Capacity Analyses. New York: Unknown Publisher, 2021.

(NREL), National Renewable Energy Laboratory. IEEE Standard 1547. New York: Unknown Publisher, 2018.

(NREL), National Renewable Energy Laboratory. The Los Angeles 100

(PNNL), Pacific Northwest National Laboratory. Human-in-the-Loop Simulation for Power Grid Control Centers. New York: Unknown Publisher, 2021.

(PNNL), Pacific Northwest National Laboratory. Grid Communications and Networking. New York: Wiley, 2020.

(RAP), Regulatory Assistance Project. Performance-Based Regulation. New York: Unknown Publisher, 2017.

(RMI), Rocky Mountain Institute. How 'Non-Wires Alternatives' Can Help Defer Costly Grid Upgrades. New York: CRC Press, 2018.

(RMI), Rocky Mountain Institute. Debunking the Utility Death Spiral. New York: Unknown Publisher, 2014.

(SEIA), Solar Energy Industries Association. Residential Solar. New York: Unknown Publisher, 2023.

(SEIA), Solar Energy Industries Association. Net Metering. New York: Unknown Publisher, 2023.

(SEPA), Smart Electric Power Alliance. The Changing Utility Business Model. New York: Walter de Gruyter GmbH Co KG, 2018.

(SEPA), Smart Electric Power Alliance. Non-Wires Alternatives: Case Studies from the Field. New York: Springer, 2020.

(SEPA), Smart Electric Power Alliance. The State of Distribution System Planning. New York: Springer Nature, 2021.

(WRI), World Resources Institute. A People-Centered Approach to Siting Clean Energy. New York: Springer, 2023.

Association, American Clean Power. Transmission Cost Allocation. New York: Createspace Independent Publishing Platform, 2021.

Bank, The World. Pricing Carbon. New York: World Bank Publications, 2023.

Council, GridWise Architecture. Transactive Energy. New York: Elsevier, 2015.

Electricity), ENTSO-E (European Network of Transmission System Operators for. TSO-DSO Platform: Final Report. New York: Unknown Publisher, 2019.

Elsberg, Marc. Blackout. New York: Sourcebooks, Inc., 2012.

Energy, U.S. Department of. Building a Better Grid Initiative. New York: Harvard University Press, 2022.

Energy, U.S. Department of. Large Power Transformer Study. New York: Nova Science Publishers, 2023.

Energy, U.S. Department of. Dynamic Line Ratings. New York: Unknown Publisher, 2022.

Energy, U.S. Department of. A Guide to Community Solar. New York: Unknown Publisher, 2022.

Energy, U.S. Department of. Microgrids. New York: CRC Press, 2021.

Energy, U.S. Department of. Advanced Distribution Management System (ADMS). New York: Unknown Publisher, 2017.

Energy, U.S. Department of. Integrated Resource Planning (IRP). New York: Unknown Publisher, 2022.

Energy, U.S. Department of. Energy Forecasting for the Smart Grid. New York: CreateSpace, 2019.

Energy, U.S. Department of. Long-Duration Storage Shot. New York: Unknown Publisher, 2021.

Energy, U.S. Department of. U.S. National Clean Hydrogen Strategy and Roadmap. New York: Unknown Publisher, 2022.

Flynn, Adam. Solarpunk: A Reference Guide. New York: Unknown Publisher, 2014.

Forum, World Economic. The rise of the 'prosumer': how active consumers are changing the energy system. New York: Dykinson, 2017.

Group, ABB. The Digital Substation: A key element of the smart grid. New York: Springer Nature, 2017.

Group, Clean Energy. Resilience for Free: How Solar+Storage Can Power Critical Facilities. New York: Unknown Publisher, 2018.

House, The White. FACT SHEET: Biden-Harris Administration Announces New Standards and Major Progress for a Made-In-America National Network of Electric Vehicle Chargers. New York: National Academies Press, 2023.

House, The White. Justice40 Initiative. New York: Unknown Publisher, 2021.

Initiative, Carbon Tracker. Stranded assets: a climate risk challenge. New York: Unknown Publisher, 2017.

Initiative, MIT Energy. The Utility of the Future. New York: Unknown Publisher, 2016.

Insights, Guidehouse. Energy as a Service. New York: Unknown Publisher, 2020.

LLP, Stoel Rives. Renewable Energy Curtailment: A-to-Z. New York: Edward Elgar Publishing, 2020.

Laboratories, Sandia National. Microgrid Controls. New York: Springer, 2018.

Laboratory, Lawrence Berkeley National. Queued Up: Characteristics of Power Plants Seeking Transmission Interconnection, 2023 Edition. New York: Unknown Publisher, 2023.

Lai, A. E. Motter and Y.-C.. Cascade-based attacks on complex networks. New York: Unknown Publisher, 2002.

Media, Greentech. Grid-Forming Inverters: A Primer on the Next-Gen of Grid-Edge Tech. New York: CRC Press, 2021.

N/A. Fictional Quote. New York: Unknown Publisher, 2024.

President, Executive Office of the. A Policy Framework for the 21st Century Grid: Enabling Our Secure Energy Future. New York: Createspace Independent Publishing Platform, 2011.

Quality, The White House Council on Environmental. Opportunities to Advance Equity in Utility Regulation: A Report to Congress. New York: Unknown Publisher, 2021.

Scientists, Union of Concerned. How Energy Storage Works. New York: Springer Nature, 2019.

Service, Congressional Research. Jurisdiction and Regulation of the U.S. Electric Grid. New York: Unknown Publisher, 2022.

Service, Congressional Research. Physical Security of the U.S. Power Grid. New York: Unknown Publisher, 2022.

Society, IEEE Power and Energy. Impact of Distributed Generation on Power System Protection. New York: Springer, 2011.

Solar, Vote. The Value of Solar. New York: Cambridge University Press, 2019.

SolarEdge. What Is a Smart Inverter?. New York: Unknown Publisher, 2021.

University, Stanford. How AI can help create a more flexible, reliable and clean electric grid. New York: Springer Nature, 2020.

Wikipedia. Virtual Power Plant. New York: CRC Press, 2023.

Wood, Elisa. Military Microgrids: A Large and Growing Market. New York: Unknown Publisher, 2021.

Jianwei Huang, et al.. The Internet of Energy: A Market-based Approach for the Smart Grid. New York: John Wiley Sons, 2011.

*Synapse traces*

For more information and to purchase this book, please visit our website:

NimbleBooks.com

*Grid Modernization: One Grid or Many?*

www.ingramcontent.com/pod-product-compliance
Lightning Source LLC
Chambersburg PA
CBHW040311170426
43195CB00020B/2933